THE SOUTHERN REGION (B R) CLASS 73 AND 74 LOCOMOTIVES

A PICTORIAL OVERVIEW

Front Cover: Class 73/1 73101 *The Royal Alex* descends from Dee Bridge (Llangollen Railway) and passes Pentre Felin as it approaches Llangollen on 11 April 2008 whilst working the 15:15 Carrog-Llangollen service.

Rear Cover: Class 73/1 73106 enters Clapham Junction yard on 11 September 1981 whilst working a London Waterloo-Clapham stock move.

THE SOUTHERN REGION (B R) CLASS 73 AND 74 LOCOMOTIVES

A PICTORIAL OVERVIEW

FRED KERR

PEN & SWORD TRANSPORT

AN IMPRINT OF PEN & SWORD BOOKS LTD.
YORKSHIRE – PHILADELPHIA

First published in Great Britain in 2023 by
Pen and Sword Transport
An imprint of
Pen & Sword Books Ltd.
Yorkshire - Philadelphia

ISBN 978 1 39904 881 1

A CIP catalogue record for this book is available from the British Library.

Typeset in 10/12 Palatino by SJmagic DESIGN SERVICES, India.

Printed and bound in India by Replika Press Pvt. Ltd.

Pen & Sword Books Ltd incorporates the imprints of Pen & Sword Books Archaeology, Atlas, Aviation, Battleground, Discovery, Family History, History, Maritime, Military, Naval, Politics, Railways, Select, Transport, True Crime, Fiction, Frontline Books, Leo Cooper, Praetorian Press, Seaforth Publishing, Wharncliffe and White Owl.

For a complete list of Pen & Sword titles please contact

PEN & SWORD BOOKS LIMITED
George House, Units 12 & 13, Beevor Street, Off Pontefract Road,
Barnsley, South Yorkshire, S71 1HN, England
E-mail: enquiries@pen-and-sword.co.uk
Website: www.pen-and-sword.co.uk

or

PEN AND SWORD BOOKS
1950 Lawrence Rd, Havertown, PA 19083, USA
E-mail: uspen-and-sword@casematepublishers.com
Website: www.penandswordbooks.com

CONTENTS

Introduction 7

Section 1: Class JA = E6001-6006; Class 73/0 = 73001-73006 9

Section 2: Class JB = E6007-6049; Class 73/1 = 73101-73142 26

Section 3: Class 73/2 = 73201-73213; 73235 66

Section 4: Class 73/9 80

Section 5: Class HB = E6101-6110; Class 74 = 74001-74010 103

INTRODUCTION

When the Southern Railway (SR) was created at the 1923 Grouping, an early decision was required regarding the future of the electrification of the passenger network, given that three different systems were in contention. These were the 630V dc third rail system of the London & South Western Railway (LSWR) which covered 57 route miles plus the 1.5 miles of the Waterloo & City line, the 6.6V ac overhead system of the London, Brighton and South Coast Railway (LBSCR) which covered 24.5 route miles with further extensions under way and the nascent 1500V dc 3rd rail/overhead system of the South Eastern & Chatham Railway (SECR) which had been approved but not initiated. After considering all three systems, the new SR Board, led by Sir Herbert Walker, decided to continue with the LSWR system as the cost (including replacement of the LBSCR system) proved to be the cheapest for both installation and operation. The decision excluded station sidings and shunting yards, hence it was accepted that a locomotive capable of working without electric power would be needed. The consequent idea of an electric locomotive fitted with a supplementary diesel engine was first mooted in 1940 but it was 1959 before a practical and suitable design was accepted.

In July 1959, approval was given for the construction of a batch of six electro-diesel locomotives designated Class JA with fleet numbers E6001-6006; they were designed to produce 1600hp using 3rd rail electric power with the bonus of providing a further 600hp from an installed diesel engine. This was achieved by using proven equipment in the guise of the electrical equipment fitted to the recently delivered 4-car BEP/CEP trainsets (later Class 411/412) and the English Electric 4SRKT diesel engine rated at 600hp at 850 rpm as fitted to the region's fleet of Diesel Electric Multiple Units (DEMU); this equipment was fitted inside a Hastings Line-gauge bodyshell to provide universal availability throughout the Southern Region. Because the equipment was standard for the Multiple Unit fleet, the order was placed with Eastleigh Carriage & Wagon Works rather than the nearby Locomotive Works.

The first locomotive, E6001, was released from the Works on 1 February 1962, made its maiden run on 5 February 1962, then ran to Brighton where it was officially inspected on 9 February 1962 before being allocated to London Stewarts Lane where the class was based until withdrawn from service in May 1993. The remaining five locomotives soon followed from Eastleigh and were immediately allocated to London Stewarts Lane and the stud of six locomotives quickly settled into their mundane duties of freight, parcels and empty stock duties with occasional appearances on the Night Ferry (London-Paris) service between London Victoria and Dover. Passenger work was limited as the fleet was too low-geared for high speed passenger service and the Electric Train Heating (ETH) equipment was only functional when the locomotive was stationary, hence the restriction on passenger duties.

The basic concept proved sufficiently successful that a further thirty locomotives, designated Class JB and numbered E6007-6036, were ordered from English Electric's Newton-le-Willows locomotive works in April 1964, followed by a further order for thirteen, numbered E6037-6049, in January 1965. The new fleet proved sufficiently successful that over time sub-classes were created to handle specific services such as the Gatwick Express service. When British Rail extended its Total Operations Performance

System (TOPS) to cover locomotive monitoring during the 1970s, the fleet was initially identified as Classes 73/0 (the six original prototypes) and 73/1 (the forty-three production series) whilst the Gatwick Express locomotives were classified Class 73/2. At privatization, the class 73/1 fleet had been allocated to the Freight Sector and were transferred to English Welsh and Scottish Railways (EWSR) when that company bought the three main freight companies (Loadhaul, Mainline Freight and Transrail Freight) in February 1996. Shortly after the takeover, EWSR undertook a fleet review, which resulted in the decision to withdraw the Class 73/1 fleet as being 'surplus to requirements'; whilst they may have been 'surplus to requirements' of EWSR, heritage lines and new Train Operating Companies created following privatization thought them sufficiently valuable that many were purchased for further service.

The initiation of the 1967 Bournemouth Electrification Scheme identified that the Class 73 fleet was under-powered for the haulage of both Southampton Docks Boat Trains and the Channel Islands Boat Train (London Waterloo-Weymouth service) and sought a more powerful design. Noting that the Class 71 electric locomotives, rated at 2550hp, were being withdrawn from service, it was decided to select ten locomotives to fit with a Paxman 6YJXL engine rated at 650hp at 1500 rpm to become Class HB, numbered E6101-6110. Introduced to service in 1967, the fleet proved unreliable in service and were withdrawn in December 1977 after a short working life.

This album is dedicated to the history of both classes which provided an SR solution to an SR need and, despite EWSR finding no value in retaining them in 1997, they continue to provide valuable service in 2022 with little sign of them being declared 'surplus to requirements' any time soon. I hope that this album helps to show how useful and valued these locomotives continue to be some sixty years after E6001 appeared from Eastleigh Carriage & Wagon Works on 1 February 1962.

Fred Kerr
October 2023

SECTION 1: CLASS JA = E6001-6006; CLASS 73/0 = 73001-73006

1.1: BR SERVICE

The prototype series of six locomotives was an unusual design that was geared to the specific requirements of British Rail's Southern Region (BR(SR)) to enable a locomotive to take power from the 3rd rail network whilst on main-line duty yet be able to work over non-electrified routes by taking power from an installed diesel engine. This was achieved by fitting electrical equipment, similar to that of the recently-built 4BEP/4CEP trainsets supplied from Eastleigh Carriage & Wagon Works, and an English Electric 4SKRT engine, rated at 600hp at 850 rpm, as fitted to the region's fleet of Diesel Electric Multiple Units (DEMU), within a Hastings Line-gauge bodyshell.

(Image 1): 73003 poses outside the shed building at London Stewarts Lane on 4 April 1981 whilst awaiting its next duty.

The contract to build the locomotives was placed with Eastleigh Carriage & Wagon Works because the main components were stock Multiple Unit items and the first locomotive, E6001, was released from Eastleigh on 1 February 1962, ran its first train on 5 February 1962 and was revealed to the region's management at Brighton on 9 February 1962 before being allocated to London Stewarts Lane depot where the six prototype locomotives were based throughout their working life; the other five locomotives quickly followed to Stewarts Lane during 1962

The locomotives proved successful but were too lowly-geared for express passenger duty, hence quickly settled into a wide range of duties including newspaper/mail services, infrastructure services and the occasional appearance on the overnight Night Ferry (London-Paris) service between London Victoria and Dover. A further restriction on passenger duty arose from the availability of ETH which only operated when the locomotive(s) were stationary.

(Image 2): 73004 poses with Class 73/1 73126 in the depot yard at London Stewarts Lane on 21 April 1981.

(Image 3): 73002 undergoes maintenance at London Stewarts Lane depot on 14 July 1979.

(Image 4): 73005 pilots de-icing unit 005, converted from a Class 405 4-SUB 4-car trainset, as they depart from a break at Canterbury East on 4 November 1978 whilst working a Dover-Ashford de-icing service.

(Image 5): 73003 weaves through Clapham Junction on 11 September 1981 whilst working an Acton-Tolworth Coal Concentration Depot coal service.

(Image 6): 73004 calls at East Croydon on 25 September 1980 whilst working a London Bridge-Brighton mail/parcels service.

1.2: MERSEYRAIL SERVICE

The Class 73/0 fleet was quickly superseded by the Class 73/1 fleet (see Section 2) production series and were transferred to infrastructure duties but by 1993 the availability of replacement Class 37 locomotives saw the fleet being placed into store on 16 May 1993 and being officially withdrawn from service the following day. This became known to the staff at Merseyrail, then part of Regional Railways North West, who were seeking replacements for the battery locomotives that had been converted from ex Euston-Watford Class 501 Motor Open Brake Second vehicles and identified that the electro-diesel Class 73/0, with its diesel engine for non-electric duties would be a suitable replacement. Regional Railways created a new pool code (RCRB) on 2 June 1993, then arranged to transfer 73001/002/005 to Chester where they were inspected on 9 June 1993. 73001 undertook a gauging trial over the Merseyrail Wirral Lines network on 12 July 1993 then moved to Birkenhead Depot on 17 August 1993; it immediately entered the depot workshops for conversion to Merseyrail specification and was finally released to traffic during January 1994 bearing the brown/yellow livery of Merseyrail – its new operator.

E6003/73003 had been sold into preservation (see Section 1.3) with the Lavender Line and 73002/005 were delivered to Birkenhead North depot hence the remaining two class members were transferred to Birkenhead North with 73004 arriving by road to serve as a spares donor.

(Image 7): 73006 stables in Hall Road maintenance depot on 13 October 1994 showing the Merseyrail brown/yellow livery and Regional Railways branding. The Merseyrail network was initially franchised as part of Regional Railways North West franchise but became a stand-alone franchise controlled by Merseytravel in 2002.

(Image 8): 73001 stables in Hall Road maintenance depot on 21 March 1994 displaying its Merseyrail livery and Regional Railways branding.

(Image 9): 73002 is stored as a spares donor at Hall Road maintenance depot on 7 June 1995.

(Image 10): 73004 is stabled outside Birkenhead North depot as a spares donor on 29 June 1995.

(Image 11): 73005 approaches Southport Portland St crossing on 9 April 1997 whilst engaged on driver training runs between Hall Road and Southport.

(Image 12): 73005 stables outside Birkenhead North depot on 26 July 2001 in the company of Class 73/9 73906 and 73901 (see Section 4.1) as they await a decision regarding their disposal.

(Image 13): 73005 stables in Hall Road maintenance depot on 22 May 1997 in the company of Class 73/9 73906 (see Section 4.1) and 73002.

Left (Image 14): 73005 awaits departure from Birkenhead North depot on 23 April 2000 to video the Wirral Lines network as part of a route learning programme.

Below: (Image 15): 73005 approaches Hooton on 23 April 2000 whilst engaged in videoing the Ellesmere Port-Hooton line as part of a route learning programme.

(Image 16): 73006 weaves off the electric network at Southport on 12 September 1995 whilst transferring Class 508 trainsets 508113/101 from Merseyrail to Bradford Sidings for transfer to storage at MOD Kinnerton; the units were subsequently leased to Connex South East to become 508201/206 respectively.

(Image 17): 73006 powers past Bradford Sidings with the Class 508 units during the transfer move.

(Image 18): 73001 is displayed in Birkenhead North depot yard on 29 June 1995 between Class 501 vehicles ADB 977348 (ex M75180) and 977347 (ex M61180) to exhibit the new de-icing/sandite trainset. The use of the Class 501 vehicles was to avoid the need to create front end exits from the locomotive cabs as required by Merseyrail's tunnel emergency regulations.

(Image 19): 73005 stables outside Birkenhead North depot on 4 September 2002 in the company of Class 73/9 73906/901 (see Section 4.1) as they await a decision regarding their disposal.

(Image 20): 73006 arrives at Hall Road on 9 March 1995 whilst transferring 73005 and 73002 from Kirkdale Electric Multiple Unit (EMU) depot to Hall Road maintenance depot for storage.

1.3: PRESERVATION

Four members of the class entered preservation; E6003 was preserved by the Electro-Diesel Group in 1996 initially based at the Lavender Line and 73001/005/006 were preserved by the Dean Forest Diesel Association (DFDA) in September 2002 whilst 73002 was purchased as a source of spares – initially by the DFDA then in December 2021 by Locomotive Services Ltd – for 73001 which is now owned by that company.

(Image 21): E6003 *Sir Herbert Walker* stables in the diesel siding at Loughborough Central (Great Central Railway) on 15 August 2005 and …

(Image 22): … displays its original livery and crest.

(Image 23): 73006 stables beside Class 73/1 73109 *Battle of Britain 50th Anniversary* in Eastleigh Workshops Yard on 25 May 2009 during the Eastleigh Works Open Day that celebrated the site's centenary of railway activity.

Right (Image 24): 73006 stables in Eastleigh Workshops yard on the same day.

Below (Image 25): E6005/73005 stables above the car park at Bridgnorth (Severn Valley Railway (SVR)) on 24 September 2005.

(Image 26): E6005/73005 stables in the headshunt siding at Bridgnorth (SVR) on 9 March 2005.

(Image 27): E6005/73005 stables in the headshunt siding at Bridgnorth (SVR) on 24 September 2005.

Left (Image 28): 73002 stables as a spares donor in Lydney station yard (Dean Forest Railway (DFR)) on 29 February 2004.

Below (Image 29): 73006 eases into Bewdley (SVR) on 8 October 2009 whilst working the 10:25 Kidderminster-Bewdley shuttle service.

Below (Image 30): 73001 pilots Class 20 20020 as they depart from Rawtenstall (East Lancashire Railway (ELR)) on 4 July 2015 whilst working the 14:19 Rawtenstall-Heywood service during the line's annual Diesel Gala event.

(Image 31): 73001, working in top 'n' tail mode with Class 37/0 37109, eases out of Ramsbottom Sidings (ELR) on 18 February 2017 whilst hauling the stock to form a Ramsbottom-Bury Bolton St shuttle service and …

(Image 32): … waits for the signal to enter Ramsbottom station.

SECTION 2: CLASS JB = E6007-6049; CLASS 73/1 = 73101-73142

The six prototype locomotives proved sufficiently successful that an order was placed with English Electric in April 1964 for a further thirty locomotives (E6007-6036/73101-129) with yet a further thirteen (E6037-6049/73130-73142) ordered in January 1965; the order was supplied from English Electric's Newton-le-Willows site. The production series was built to a revised specification which included the locomotives to be equivalent, on electric power, to the Class 33 locomotives rated at 1550hp; this was met by the electric portion providing 1600hp. Further differences from the prototypes included the gearing of the locomotives for express passenger duties and a change to the ETH which would henceforth operate whilst working services rather than when stationary as with the prototypes. Although capable of working in multiple with the region's diesel and electric multiple units, the differences meant that the Class 73/1 could not work in multiple with Class 73/0 locomotives.

2.1: LOCOMOTIVES ON SHED

(Image 33): 73118 stables in the yard of Eastleigh depot on 5 December 1976.

(Image 34): 73128 stables in the yard of London Hither Green depot on 8 January 1978

(Image 35): 73103 and 73111 stable in the depot yard of London Stewarts Lane on 25 April 1981.

(Image 36): 73102 receives maintenance inside London Stewarts Lane depot on 25 April 1981.

(Image 37): 73105 carries large logo blue livery on 21 January 1984 as it stables in Basingstoke station yard awaiting its next duty.

(Image 38): *73109 Battle of Britain 50th Anniversary* carries Network South East livery on 16 October 1994 when it appeared at Crewe International Electric Depot as a 'guest' locomotive and display item when the depot held an Open Day to introduce Class 92 locomotives to service.

2.2: PASSENGER SERVICES

The class members were often to be found on overnight mail/newspaper services, especially those with passenger accommodation which operated from London stations to cater for theatre/concert audiences seeking to return home to the Home Counties. Their duties included a wider range of services than the Class 73/0 locomotives hence proving to be useful motive power in the varied workings of the extensive BR(SR) network of electrified routes.

(Image 39): 73121 awaits departure from London Waterloo in the early hours of 24 June 1983 whilst working the 02:45 Waterloo-Poole newspaper service which included passenger accommodation for late evening passengers.

(Image 40): 73101 *Brighton Evening Argus* awaits departure from London Waterloo on the morning of 25 June 1983 whilst working a Waterloo-Portsmouth newspaper/mail service.

(Image 41): 73105 awaits departure from London Waterloo in the early hours of 8 October 1977 whilst working the 02:15 Waterloo-Bournemouth newspaper/mail service which included passenger accommodation.

(Image 42): In this image, 73125 awaits departure for the same service, but on a summer's night, 24 June 1983.

(Image 43): 73142 *Broadlands* propels an ecs consist, comprising Class 430 trainset 420 and Class 423 trainset 7828, into Eastleigh whilst working a Bournemouth-Eastleigh stock move. Of interest is that the trainset is driven from the leading (unpowered) Class 430 driving cab. The locomotive was named in commemoration of its involvement with the Royal Train which carried the Prince and Princess of Wales to Broadlands at nearby Romsey at the start of their honeymoon after their wedding on 29 July 1981.

(Image 44): 73141 powers past Railway Halt (the platform for staff employed at Wimbledon and Wimbledon Park EMU depots) on 21 May 1980 whilst working the Channel Islands Boat Train from Weymouth to Waterloo.

(Image 45): 73141 powers through St Denys on October 28 1977 whilst working the 08:15 Totton-Eastleigh commuter service that utilised the stock of the early morning 02:15 London Waterloo-Bournemouth newspaper/passenger service.

(Image 46): 73123 shunts the stock of a boat train in Southampton Docks on 29 July 1979 in preparation for taking passengers from RMS *Queen Elizabeth 2* to London Waterloo.

The Night Ferry service between Paris and London Victoria was a regular duty for the class, working between Victoria and Dover. Scenes of the overnight working from Paris include …

(Image 47): … 73110 standing after arriving in Victoria on 8 July 1979.

Left (Image 48): … 73140 approaching the platform at Victoria on 22 July 1979 then …

Below (Image 49): … standing at the platform as passengers detrain.

(Image 50): 73142 sees royal duty on 12 June 1979 as it eases out of Gatwick Airport whilst conveying a royal guest to London Victoria as part of a State Visit to the United Kingdom.

Opposite (Image 51): 73106 enters Clapham Junction yard on 11 September 1981 whilst working a London Waterloo-Clapham stock move.

2.3: FREIGHT SERVICES

Class members undertook a wide variety of the freight services operated by the Southern Region, despite being mainly an electrified passenger network. In the twenty-first century these include an increasing number of services operated on behalf of Network Rail, as successor to Railtrack (see Section 4.3)

Above (Image 52): 73134 awaits departure from London Victoria 25 June 1983 whilst working a early morning Victoria-Dover newspaper/mail service.

Opposite above (Image 53): 73112 awaits departure from London Waterloo in the early hours of 17 December 1981 whilst working a Waterloo-Portsmouth newspaper/mail service.

Opposite below (Image 54): 73139 stables in London Waterloo on 17 March 1981 awaiting its next duty.

72
73112

73139

(Image 55): 73111 and 73114 pass Clapham Junction on 17 December 1981 whilst transferring Class 302 trainset 302294 as a stock move from Eastleigh Carriage and Wagon Works to Ilford Carriage Sidings following overhaul then …

(Image 56): … curve onto the North London line to gain access to the Great Eastern line at Stratford.

(Image 57): 73133 awaits departure from London Victoria in the early hours of 25 June 1983 whilst working a Victoria-Ramsgate newspaper/mail service.

Above (Image 58): 73127 passes St Denys on 28 October 1977 whilst working a Bournemouth-Eastleigh van service that would later move to Clapham Junction yard for overnight newspaper/mail services from London Waterloo.

Left (Image 59): 73138 passes St Denys on 28 October 1977 whilst working a Southampton Maritime-Eastleigh service of container flat wagons for repair/overhaul at the workshops.

(Image 60): 73138 calls at Eastleigh for a crew change on 22 January 1983 whilst working a Northam-Hoo Junction cement service.

(Image 61): 73142 *Broadlands* and 73131 stable in Eastleigh station yard on 23 January 1983 whilst awaiting departure with an Eastleigh-Clapham Junction yard stock move …

(Image 62): … and remained until later in the evening waiting to depart for Clapham Junction.

(Image 63): 73137 weaves through Clapham Junction on 2 April 1981 whilst working an Acton-Tolworth Coal Concentration Depot coal service.

(Image 64): 73101 *Brighton Evening Argus* pilots 73141 as they pass Putney on 16 September 1981 whilst working the Cricklewood-Northfleet leg of a Toton-Northfleet Cement Works MGR coal service.

2.4: PRIVATISATION

When British Rail was privatised on 1 April 1994, the Class 73/1 fleet had been allocated to the freight companies hence, when these were bought by EWSR, a subsequent traction review decided that the fleet should be withdrawn from service. The fleet was subsequently offered for sale leading to class members being bought by preservation groups as well as railway companies including Eurostar (which bought and modified 73118/130 for use as rescue locomotives), FM Rail/Fragonset, Gatwick Express (Class 73/2 fleet), GB Railfreight, Railtrack/Network Rail, RT Rail (which bought 73107 when FM Rail entered administration) and South West Trains (SWT) which retained 73109 *Battle of Britain 50th Anniversary* as a 'Thunderbird' locomotive. Over time, GB Railfreight subsequently found them to be useful locomotives so bought as many as possible during this century from both preservation bodies and other railway operators thus becoming the largest operator of these locomotives (see Sections 2.7/3.4/4.3/4.4).

Right (Image 65): 73109 *Battle of Britain 50th Anniversary* bears early Stagecoach corporate livery as it races through Petersfield on 7 October 2004 whilst undertaking route-learning duties between Woking and Fratton.

Below (Image 66): 73109 *Battle of Britain 50th Anniversary* bears the house colours of Stagecoach-owned SWT as it poses in Eastleigh Works Yard as a display item on 25 May 2009 during a Works Open Day to celebrate the site's centenary as a locomotive works facility.

Whilst most purchasers of Class 73/1 locomotives were preservation groups, among the commercial operators Eurostar purchased 73118/130 through its Euro Passenger Services arm to operate as rescue locomotives over its HS1 network system pending completion of the route to St Pancras; the duo was fitted with Scharfenberg couplers to enable this task.

Left (Image 67): Details of the Scharfenberg couplers when in use as displayed when 73118 and 73130 were coupled together at London Stewarts Lane depot shortly after fitting during the 1990s. *(Courtesy of Brian Morrison)*

Below (Image 68): View of 73130 on display at London Stewarts Lane depot shortly after the fitting of the Scharfenberg coupling in the 1990s. *(Courtesy of Brian Morrison)*

(Image 69): 73118 passes Tonbridge West on 30 July 1996 whilst moving barrier support vehicles from Dollands Moor to North Pole Depot. *(Courtesy of Brian Morrison)*

(Image 70): 73130 passes Bromley South on 2 March 2004 whilst making a locomotive move from North Pole depot to Dollands Moor. *(Courtesy of Brian Morrison)*

Aware of their value a pair of Class 73/1 locomotives were bought by Harry Needles Railway Company and stored at his Barrow Hill base pending re-sale or breaking up for spares.

Above (Image 71): A visit to the site on July 12 2003 found, among the assembled stored locomotives, 73119 and 73138 awaiting sale/disposal.

Left (Image 72): 73138 is here awaiting sale/disposal.

2.5: PRESERVATION

Although the Class 73/1 fleet combined both diesel and electric traction, the diesel engine proved more attractive to heritage lines as a source of traction, given that its 600hp diesel engine was more powerful than the 400hp of preserved Class 08 or smaller shunting locomotives then being operated on heritage lines whilst the availability of ETH made them useful for passenger duty on winter days.

F… 1994 privatisation of British b… ong the classes deemed to t… h the ELR in September 1999 l… ses, a pair of Class 73/1 … med the nucleus of the event.

… *The Burma Star* and resident … on 12 September 1999 whilst …

(Image 74):
73133 *Bluebell Railway* pilots 73129 *City of Winchester* out of Ramsbottom (ELR) on 12 September 1999 whilst working the 13:45 Rawtenstall-Bury Bolton St service.

Right (Image 75):
73133 *Bluebell Railway* pilots 73129 *City of Winchester* as they call at Summerseat (ELR) on 10 September 1999 whilst working the same service.

Below (Image 76):
73133 *Bluebell Railway* pilots Class 35 'Hymek' D7076 and Class 33/2 33202 *The Burma Star* as they power through Burrs on 12 September whilst working the 16:15 Rawtenstall-Bury Bolton St service.

(Image 77):
73133 *Bluebell Railway* eases the stock of the 11:02 Ramsbottom-Bury Bolton St shuttle service out of the sidings at Ramsbottom (ELR) on 10 September 1999 and ...

Left (Image 78):
... awaits to depart from Ramsbottom with 73129 *City of Winchester* and Class 31/1 31110 *Traction Magazine* forming the rear of the shuttle trainset.

Below (Image 79):
73133 *Bluebell Railway* pilots 73129 *City of Winchester* as they curve out of Rawtenstall (ELR) on 10 September 1999 whilst working the 13:45 Rawtenstall-Bury Bolton St service.

In July 2007 Barrow Hill held a Diesel Gala event during which resident 73138 powered a number of shuttle services, working in top 'n' tail mode with various locomotives. On 8 July 2007 these workings included 73138 working in top 'n' tail mode with …

(Image 80): … Class 06 06003 as it entered the display yard.

(Image 81): … Class 37/0 37178 as it approached the display yard.

(Image 82): … Class 37/0 37178 as it set off from the branch terminal point to return to the display yard.

(Image 83):
73109 *Battle of Britain 50th Anniversary* poses beside preserved Class 73/0 73006 in Eastleigh Works yard on 25 May 2009 whilst serving as display items during the Works Open Day to celebrate 100 years of railway locomotive building and maintenance.

Left (Image 84):
E6043/73136 *Perseverance* poses inside Eastleigh Works on 25 May 2009 whilst appearing as a display item during the Works Open Day to celebrate the site's centenary.

Below: (Image 85): 73133, devoid of its *Bluebell Railway* nameplates, stables in Barry Sidings (Vale of Glamorgan Railway/Barry Tourist Railway) on 7 July 2005 whilst awaiting restoration.

(Image 86): 73138 is the rear of the Barrow Hill shuttle service powered by Class 20 D8132 as it departs from the display yard on 8 July 2007 to head up the short branch.

Right (Image 87): 73134 *Woking Homes 1885-1985* stables in Harry Needles' storage yard on 4 April 2009 whilst awaiting a decision as to its future.

Below (Image 88): E6047/73140 stables in the yard at Tunbridge Wells (Spa Valley Railway) on 17 June 2011 whilst awaiting its next duty.

E6036/73129 has been bought by the Gloucestershire Warwickshire Railway where it is maintained at the railway's base at Toddington.

Views of the locomotive at the railway include it …

Right (Image 89): stabling on Toddington shed on 19 August 2015 and …

Below (Image 90)/ Bottom (Image 91): … stabled on the shed yard at Toddington on 29 August 2016 as viewed from a passing train.

(Image 92): In its final years of service, 73101 *The Royal Alex* was re-liveried into Pullman livery and named *The Royal Alex* to support the VSOE Pullman service; this livery was retained when the locomotive was preserved by the Dean Forest Diesel Association. 73101 *The Royal Alex* is seen stabled in Lydney Yard (DFR) on 6 October 2004 whilst it awaited restoration to working order.

(Image 93): 73101 *The Royal Alex* was the 'guest locomotive' at the Llangollen Railway on 11 April 2008 when the line ran a Diesel Gala weekend; views of the locomotive include it reversing onto the rear of the 10:00 Llangollen-Carrog service to form a top 'n' tail consist with Class 37/9 37901 *Mirrlees Pioneer* and …

(Image 94): … curving out of Glyndyfrdwy whilst working the 13:00 Llangollen-Carrog service.

(Image 95): 73101 *The Royal Alex* approaches Llangollen (Llangollen Railway) on 11 April 2008 whilst working the 10:45 Carrog-Llangollen service.

(Image 96): 73101 *The Royal Alex* stables at Carrog (Llangollen Railway) on 11 April 2008 whilst awaiting its next duty.

(Image 97): 73101 *The Royal Alex* descends from Dee Bridge and passes Pentre Felin on 11 April 2008 as it approaches Llangollen (Llangollen Railway) whilst working the 15:15 Carrog-Llangollen service.

(Image 98): 73101 *The Royal Alex* stables in Kidderminster shed yard (SVR) on March 23 2008 whilst awaiting its next duty during a Diesel Gala event when it was a 'guest' locomotive.

2.6: MERSEYRAIL/NETWORK RAIL SERVICES

The Merseyrail network, including the 3rd rail electric lines, is regularly visited by Network Rail Inspection Trains which employed Class 73/1 locomotives for use over the electrified lines as part of its programme of inspections until these were replaced by Class 73/9 traction (see Section 4). These services operate overnight when few other trains are operated but security precautions, including the locking of stations overnight, makes photography difficult, hence these images taken at Southport are now taken at Liverpool Lime St during a lengthy break there (see Section 4).

(Image 99): 73138, working in top 'n tail mode with 73107 *Redhill 1844 – 1944*, stands in Southport in the early hours of 14 February 2012 after arriving with the Wigan Springs Branch – Southport leg of the inspection service.

(Image 100): 73107 *Redhill 1844-1994*, working in top 'n' tail mode with 73138, stands after arriving in Southport in the early hours of 9 March 2013 with the Liverpool Tuebrook Sidings-Southport leg of the inspection service.

(Image 101): 73107 *Redhill 1844-1994*, working in top 'n' tail mode with 73138, awaits departure from Southport in the early hours of 14 February 2012 whilst working the Southport-Wigan Wallgate leg of the inspection service.

(Image 102): 73138, working in top 'n' tail mode with 73107 *Redhill 1844-1994*, stables in Kirkdale Traction and Rolling Stock Maintenance Depot (T&RSMD) on 13 February 2012 whist awaiting an evening start to its tour of inspection of the Merseyrail Northern Line electric network.

(Image 103): 73138, working in top 'n' tail mode with Class 73/2 73201 *Broadlands,* stands in Southport early on 29 March 2014 after working the Crewe-Southport leg of the inspection service.

(Image 104): 73107 *Redhill 1844-1994*, working in top 'n' tail mode with 73138, awaits departure from Southport early on 14 February 2012 whilst working the Southport-Wigan Wallgate leg of the inspection service.

(Image 105): 73138, working in top 'n' tail mode with 73107 *Redhill 1844-1994*, awaits departure from Southport in the early hours of 9 March 2013 whilst working the Southport-Ormskirk leg of the inspection service.

2.7: GB RAILFREIGHT

GB Railfreight is a freight company that began operating after privatisation and, as part of its expansion, bought six redundant Gatwick Express Class 73/2 locomotives (73203-73207/209) in 2004 (see Section 3). Such was their value to the company that it bought further examples as and when they became available, becoming in January 2022 the largest operator of the class in its various forms. A point of interest is that some class members carry the names of office staff or people related to either the company or the locomotives' histories. The fleet is based at Tonbridge from where locomotives operate a range of services on behalf of Network Rail within the ex-BR(SR) network of lines.

(Image 106): 73128 *O.V. Bulleid CBE*, working in top 'n' tail mode with 73109 *Battle of Britain 50th Anniversary*, stables with a Rail Head Treatment Train in Tonbridge West yard on 15 October 2018 whilst awaiting its next tour of duty.

(Image 107): 73107 *Tracy* stables in Tonbridge West yard on 15 October 2018 whilst awaiting its next duty.

GB Railfreight occasionally provides 'guest' locomotives to heritage lines and has developed a relationship with the SVR in recent times. A sign of this relationship was noted during the SVR's Diesel Gala in May 2018 when 73136 *Mhairi* was provided as 'guest' locomotive: scenes of 73136 Mhairi at work on 18 May include it …

(Image 108): … departing from Kidderminster whilst working the 15:49 Kidderminster-Bewdley shuttle service and …

(Image 109): … being piloted by Class 37/5 37688 as the duo depart from Kidderminster whilst working the 18:25 Kidderminster-Bridgnorth service.

The company staff organises a bi-annual multi-day railtour for charity and operates company traction on unusual routes as part of the attraction. In September 2018, its 'Out of the Ordinary' charity railtour resulted in Class 73/1 locomotives operating in the north of England and Scotland. In the early hours of 22 September 2018, 73141 *Charlotte* and 73107 *Tracy* worked in top 'n' tail mode with Class 92 92033 and were noted at Blackpool North waiting to lead the Blackpool North-Preston leg of the Blackpool North-Mossend portion of the railtour.

Above (Image 110): 73141 *Charlotte* and 73107 *Tracy* await departure from Blackpool North.

Below left (Image 111)/Below right (Image 112): 73107 *Tracy* is piloted by 73141 *Charlotte* as they await departure from Blackpool North.

(Image 113): The southbound working the following day saw the pair of Class 73/1 locomotives forming the rear of the train as it passed Euxton with the Carlisle-Crewe leg of the railtour hauled by Class 86/6 86637 and 86639 with 73141 *Charlotte* and 73107 *Tracy* on the rear and …

(Image 114): … 73107 *Tracy* and 73141 *Charlotte* on the rear.

SECTION 3: CLASS 73/2 = 73201-73213; 73235

The Class 73/2 sub-class was created in early 1988, using modified Class 73/1 locomotives to power the dedicated Gatwick Express service between London Victoria and Gatwick Airport that had been introduced in May 1984. The initial operation comprised MkIIf coaches with a Luggage Van, modified from a Class 414 Driving Motor Brake Second at one end and a Class 73/1 locomotive providing power at the other end. The initial modified Class 73/1 locomotives suffered problems coping with the high speeds of the service which led to the decision to create a dedicated pool of twelve locomotives (later increased to thirteen) that were specially modified for the service – including re-gearing to handle the higher speeds. This dedicated pool, comprising 73142/137/127/125/124/123/122/12 1/120/116/113/102/was renumbered to 73201-212 during 1989 with 73135 becoming 73235 in April 1991 then 73112 becoming 73213 in July 1996. The fleet operated until 2005 as, after the privatisation of BR in April 1994, the Gatwick Express franchisee elected to order a fleet of EMU trainsets which were introduced to service in 2000. Reliability problems, however, continued to require the use of Class 73/2 locomotives as standby – a requirement that finished in 2005 when a new fleet of dedicated EMU trainsets was introduced.

3.1: GATWICK EXPRESS

(Image 115): 73204 *Stewarts Lane 1860-1985* stands in Victoria on 17 October 1990 awaiting departure whilst working a Gatwick Airport service.

(Image 116): 73206 *Gatwick Express* stables at the platform in Victoria on 17 October 1990 awaiting its next departure to Gatwick Airport.

(Image 117): 73210 *Selhurst* stables at the platform in Victoria on 17 October 1990 awaiting its next departure to Gatwick Airport.

73201 was one of the class members which remained operating the Gatwick Express services until replaced by EMU trainsets in 2005.

(Image 118): A visit to the Gatwick Airport area on 23 November 2004 noted 73201 standing at Gatwick Airport awaiting departure to London Victoria and …

(Image 119): … propelling a London Victoria-bound service out of Gatwick Airport and later …

(Image 120): … propelling a Gatwick Airport-London Victoria service through Earlswood.

3.2: PRIVATISATION

Once released from Gatwick Express duties, the fleet members were bought by train operators for continued use on the privatised network (see also Section 3.3 and Section 3.4). 73235 was obtained by SWT to act as both 'Thunderbird' locomotive and undertake depot pilot/stock move duties whilst 73202 was retained by Gatwick Express for 'Thunderbird' duties until released in 2005 and returned to Porterbrook who re-leased it to Connex South East then to South East Trains for both 'Thunderbird' and staff route learning duties.

(Image 121): 73235 poses in Wimbledon Traincare yard on 3 February 2005 whilst shunting Class 421 3-CIG trainset 1883.

(Image 122): 73235 and 73109 *Battle of Britain 50th Anniversary* pose in Bournemouth Traincare Depot on 25 February 2008 whilst undertaking shunting duties.

(Image 123): 73202 *Dave Berry* pilots Class 73/1 73136 *Perseverance* as they work in top 'n' tail mode with Class 73/1 73109 *Battle of Britain 50th Anniversary* on 23 October 2007 as they power the Class 73 Preservation Society's 11:55 London Victoria-Gillingham Queen Of Scots charter, here noted passing Shortlands on the outward journey. (*Images 121-123 courtesy Brian Morrison*)

3.3: PRESERVATION

When the Gatwick Express fleet was withdrawn from service, most survived to undertake further service; five locomotives survived to be bought by train operating companies whilst a further six were bought by GB Railfreight and 73210 entered preservation.

In July 2013, 73210 *Selhurst* was the 'guest locomotive' at the ELR's annual Diesel Gala event held each July.

(Image 124): 73210 *Selhurst* is piloted by Class 33/1 33109 *Captain Bill Smith RN* as they arrive at Ramsbottom on 7 July 2013 whilst working the 11:46 Rawtenstall-Heywood service.

Scenes from the visit of 73210 *Selhurst* to the ELR Diesel Gala event include it …

(Image 125):
… departing from Rawtenstall on 5 July 2013 whilst working the 17:06 Rawtenstall-Heywood service.

Right (Image 126):
… piloting Class 33/1 33109 *Captain Bill Smith RN* as they call at Irwell Vale on 7 July 2013 whilst working the 09:35 Heywood-Rawtenstall service.

Below (Image 127):
… crossing the River Irwell on 5 July 2013 as it departed from Summerseat whilst working the 12:24 Bury Bolton St-Ramsbottom shuttle service.

(Image 128): ... approaching Summerseat on 5 July 2013 whilst working the 13:24 Ramsbottom-Bury Bolton St shuttle service.

(Image 129): ... curving past Townsend Fold on 5 July 5 2013 whilst working the 15:45 Heywood-Rawtenstall service.

3.4: RAILTRACK/NETWORK RAIL

When released from Gatwick Rail service, 73212/213 were bought by Railtrack to service infrastructure trains over the electrified ex BR(SR) network, especially on services when power was cut off for engineering work to take place. When Railtrack entered administration, its assets, including the pair of Class 73/2s, passed to Network Rail to continue operating the same range of services.

(Image 130): 73212/213 stable in Tonbridge West yard, their nominal home 'depot' on 8 June 2005 whilst awaiting their next duty.

Right (Image 131)/Below (Image 132): 73212 visited Barrow Hill on Railtrack business in October 2001 and became the 'guest locomotive' for an Open Day event held on 7 October when it powered a brake van shuttle service in top 'n' tail mode with Class 14 D9525.

(Image 133): When passed to Network Rail the duo were re-liveried in the company's yellow livery as noted in November 2007 when they visited Barrow Hill with a de-icing trainset. During a gala event held at Barrow Hill … 73213 was noted at the rear of the consist on 10 November 2007 whilst …

(Image 134): … 73212 was noted at the head of the consist on 9 November 2007.

3.5: GB RAILFREIGHT

GB Railfreight was created after the privatisation of BR in 1994, initially as a freight operator providing services to Railtrack – later Network Rail – but it grew in size by a combination of gaining passenger licences and becoming part of larger companies. When the Class 73/2 locomotives were released from Gatwick Express duty, the company bought six locomotives (73203-207/209) with the intent of returning four to active service powering Network Rail services. The chosen locomotives were painted in corporate livery with some named after staff who worked within the company; they proved sufficiently successful that the company began buying further Class 73 locomotives (see Section 2.7) from both other companies and preservation groups willing to sell in order to fund other ventures whilst being pleased to see the locomotives return to main line activity.

A brief visit to Tonbridge West stabling point on 8 June 2005 noted examples of GB Railfreight locomotives, replete in GB Railfreight livery, stabled around the yard. Images from the visit include …

Above (Image 135)/Below (Image 136): ... 73206 *Lisa* with Railtrack-owned 73212/213 stabled on an adjacent line on 8 June 2005.

(Image 137): ... on the same day, 73204 *Janice* and 73209 *Alison* arriving to await their next duty.

Left (Image 138): ... 73206 *Lisa*, 73204 *Janice* and 73209 *Alison* stabled awaiting their next duty.

Below (Image 139): On a later visit made on October 15 2018, the yard hosted Class 73/9 locomotives (see Section 4.3) and 73201 *Broadlands*, re-liveried in BR Corporate Blue livery and logo, awaiting their next duties.

(Image 140): When 73201 *Broadlands* was bought by GB Railfreight it was repainted into BR Corporate Blue livery but had lost its name during time spent as a Gatwick Express locomotive. This was restored during Eastleigh Works Open Day held in May 2009. Now re-named, 73201 *Broadlands* poses as an Open Day exhibit in Eastleigh Works yard on 25 May 2009.

(Image 141): As part of their Network Rail duties, GB Railfreight Class 73s visit the Merseyrail electric network with a regular Inspection Train. In the early hours of 29 March 2014 73201 *Broadlands*, working in top 'n' tail mode with Class 73/1 73138, awaited departure from Southport with the Southport-Ormskirk leg of the Inspection Train.

A visit to Eastleigh Station on 13 March 2011 noted a trio of Class 73/2 locomotives stabled in the siding adjacent to the station, a regular stabling point for GB Railfreight locomotives. The trio comprised …

Above (Image 142): … 73208 *Kirsten* in BR Corporate Blue livery with logo.

Right (Image 143): … 73212 bearing a modified version of the First Group livery applied when GB Railfreight was part of First Group for a short period.

Below (Image 144): … 73213 bearing a modified version of the First Group livery and heading the line-up comprising 73212 and 73208 *Kirsten*.

SECTION 4: CLASS 73/9

The Class 73/9 sub-group has, over time, been allocated to four separate batches of locomotives – namely:

(i) 73901/906 allocated by Merseyrail to its modified Class 73/0 locomotives
(ii) 73951/952 allocated to LORAM of Derby to a pair of Class 73/1 locomotives rebuilt and dedicated to Network Rail services
(iii) 73961-73965 allocated to GB Railfreight locomotives rebuilt for dedicated Network Rail services
(iv) 73966-73971 allocated to GB Railfreight for locomotives rebuilt for dedicated Caledonian Sleeper services within Scotland – normally north of Edinburgh.

4.1: MERSEYRAIL = 73901/73906

When Merseyrail obtained its Class 73/0 locomotives, it was creating a new de-icing/sandite train and a prototype trainset was introduced in June 1995 (see Image 18). Further modifications were made to the locomotive, hence its reclassification to 73/9, and the subsequent modifications made to Class 73/0 73006 as a back-up resulted in that locomotive also being re-classified and re-numbered.

(Image 145): Class members rarely moved off the Merseyrail electric network but on 29 July 1996, a works train was required to move from Liverpool Tuebrook Sidings to Chester with a consist that was out of gauge for transfer via the Merseyrail tunnels. It was therefore routed via Earlestown and the West Coast Main Line, hence the sight of 73901 passing Winwick with the works train en route to Chester.

(Image 146): 73901, formed between a pair of ex Class 501 vehicles as a de-icing/sandite trainset, stables in Hall Road maintenance depot on 6 May 1996 with Class 73/0 73005 and 73002 stabled behind it.

(Image 147): 73901 stands in Southport station on 3 September 1996 whilst returning to Kirkdale T&RSMD after delivering the preserved Class 503 trainset to Steamport Museum for secure storage.

Above (Image 148): 73906 stables between Class 73/0 73002 and 73005 at Hall Road Maintenance depot on 4 March 1997.

Left (Image 149): 73901 stables outside Birkenhead North traction maintenance depot on 23 April 2001 awaiting a decision as to its future.

(Image 150): 73906 stables at Hall Road Maintenance Depot on 4 March 1997.

(Image 151): 73901 heads 73906 and Class 73/0 73005 as they stable at Birkenhead North traction maintenance depot on 4 September 2002 awaiting a decision as to their future …

(Image 152): … and where on that date David Hurd and his wife Yvonne inspected the locomotives before buying them on behalf of the Dean Forest Diesel Association. David later admitted buying them for his wife as a wedding anniversary present because of her reported liking for them.

(Image 153): 73901 stables in the station yard at Lydney (DFR) on 29 February 2004 whilst awaiting restoration.

(Image 154): 73906, with 73901 in the background, stables in the station yard at Lydney (DFR) on 29 February 2004 whilst awaiting restoration.

4.2: NETWORK RAIL/RAIL VEHICLE ENGINEERING LTD = 73951-73952

In June 2014, Network Rail introduced two new locomotives that had been created by Rail Vehicle Engineering Ltd (RVEL), based at Derby, for use on infrastructure services. The duo, previously Class 73/1 73104 and Class 73/2 73211 now renumbered 73951/2 respectively, had been gutted internally then rebuilt with a pair of Cummins QSK19 engines rated at 750hp to replace the single English Electric engine rated at 600hp thus uprating the locomotive to 1500hp on diesel power whilst retaining its 1600hp on electric power. As at December 2021 the duo are based at Derby RTC and used on Network Rail services as required.

Above (Image 155):
73951 *Malcolm Brinded* poses as a display item in the Kidderminster Diesel Depot (SVR) on 21 May 2016 after failing whilst attending a diesel gala event as a 'guest locomotive'.

Right (Image 156):
73951 *Malcolm Brinded* poses as a display item at the Crewe base of Locomotive Services Ltd on 8 June 2019 during the site's Open Day event.

(Image 157): *73952 Janis Kong* pilots Class 20 D8188 into Bewdley (SVR) on 8 June 2016 whilst working the 13:47 Kidderminster-Bewdley service during its visit as 'guest locomotive' during the SVR's Diesel Gala event.

(Image 158): 73952 *Janis Kong* poses as a display item at the Crewe base of Locomotive Services Ltd on 8 June 2019 during the site's Open Day event.

(Image 159): 73951 *Malcolm Brinded* stables at the platform of Liverpool Lime St on 21 December 2021 after working the Chester-Liverpool Lime St leg of the regular Inspection Train whilst …

(Image 160): … at the other end, 73952 *Janis Kong* awaits departure with the Liverpool Lime St-Southport leg of the inspection programme.

4.3: NETWORK RAIL/GB RAILFREIGHT = 73961-73965

When Network Rail specified the creation of the RVEL Class 73/9, it also contracted GB Railfreight to provide similar Class 73/9 locomotives for use on ex BR(SR) routes. GB Railfreight preferred to replace the diesel engine with a single MTU 8V4000 R43L engine rated at 1600hp and upgrade the electric equipment; this work was contracted to Wabtec (previously Brush Traction) at Loughborough. Work began in 2013 as Class 73/2 73209/204/206/205/208 were converted to 73961-73965 respectively. The locomotives are nominally based at Tonbridge West for service over the BR(SR) electrified lines but are also occasionally operated on Network Rail services in other areas.

Shortly after their introduction to service, the ETH equipment was tested between Crewe and Warrington on 29 July 2015. The test train was operated in top 'n' tail mode with 73961 *Alison* and 73963 *Janice* and was photographed at Acton Bridge.

(Image 161): 73961 *Alison* led the outward journey from Crewe ...

Above (Image 162): … with 73963 *Janice* at the rear.

Right (Image 163): 73963 *Janice* led the return journey to Crewe …

Below (Image 164): … with *73961* Alison at the rear.

A visit was made to Tonbridge West on 15 October 2018 when …

(Image 165): … 73962 *Dick Mabbutt* was stabled in the yard.

Right (Image 166): … 73964 *Jeanette* was stabled in the yard and …

Below (Image 167): … shortly after formed the rear of the Tonbridge Yard-Brighton Network Rail Inspection Train worked in top 'n' tail mode led by 73965.

Network Rail operates regular inspection services in the Manchester/Liverpool area which include services operating over the Merseyrail 3rd rail electric network and for which the Class 73/9 locomotives are preferred in order to pass through the tunnelled areas. Although running overnight when Merseyrail stations are closed/in darkness, the schedule includes a respite at Liverpool Lime St between 23:40-00:04 for crew changes and needs break.

An inspection train visiting Liverpool Lime St. on 27 June 2016 was ...

Right (Image 168):
... noted with 73963 *Janice* standing after arriving with the Chester-Liverpool Lime St leg ...

Below (Image 169):
... whilst 73962 *Dick Mabbutt* stood at the other end awaiting departure with the Liverpool Lime St-Southport leg of the programme.

An inspection train was scheduled to operate in June 2016 and a visit to Liverpool Lime St on the 27th noted …

Above (Image 170)/ Left (Image 171): … 73963 *Janice* standing after arriving with the Chester-Liverpool Lime St leg …

(Image 172): … whilst 73962 *Dick Mabbutt* stood at the other end awaiting departure with the Liverpool Lime St-Southport leg of the programme.

An inspection train was scheduled to operate in October 2016 and a visit to Liverpool Lime St on 25 October 2016 noted …

Above left (Image 173)/Above right (Image 174): … 73965 standing after arriving with the Chester-Liverpool Lime St leg of the programme …

Above (Image 175)/ Right (Image 176): … whilst at the other end 73963 *Janice* awaited departure with the Liverpool Lime St-Southport leg of the programme.

An inspection train was scheduled to operate in January 2017 and a visit to Liverpool Lime St on 17 January noted …

Above (Image 177)/ Left (Image 178): … 73962 *Dick Mabbutt* standing after arriving with the Chester-Liverpool Lime St leg of the programme …

(Image 179): … whilst 73964 *Jeanette* awaited departure with the Liverpool Lime St-Southport leg of the programme.

An inspection train was scheduled to operate in September 2021 and a visit to Liverpool Lime St on 28 September 2021 noted …

Above left (Image 180)/Above right (Image 181): … 73964 *Jeanette* standing after arrival with the Chester-Liverpool Lime St leg of the programme …

Above (Image 182)/Right (Image 183): … whilst at the other end 73962 *Dick Mabbutt* awaits departure with the Liverpool Lime St-Southport leg of the programme.

4.4: GB RAILFREIGHT/CALEDONIAN SLEEPERS = 73966-73971

In 2015, the Caledonian Sleeper service became a franchise for which GB Railfreight tendered, and won, the contract to supply the motive power using its Class 92 fleet for the main run to Edinburgh and Glasgow then Class 73/9 traction north of Edinburgh to Aberdeen, Fort William and Inverness. This contract required six additional locomotives, 73966-73971, and these were obtained by modifying Class 73/0 73005/006 (both bought from the Dean Forest Diesel Association)/ Class 73/1 73117/105/103 and Class 73/2 73207 to become 73966-73971 respectively.

As with the initial rebuilds, 73961-73965, the original locomotives were gutted but these rebuilds had larger fuel tanks, uprated heating index and the removal of collector shoes although retaining the internal power equipment. The new locomotives began their service on 1 April 2015, with their normal usage intended to be a single locomotive working the Aberdeen and Fort William portions and a pair working the Inverness portion forward from Edinburgh.

A visit was made to Edinburgh Waverley in the early hours of 22 July 2016 when …

(Image 184): … 73966 (rebuilt from Class 73/0 73005) piloted 73968 as they awaited departure with the Fort William portion of the northbound Highland Sleeper service from London Euston.

In the early days of the operation the Class 73/9 fleet suffered alternator problems which were alleviated by the use of a Class 66/7 locomotive to provide haulage and the Class 73/9 to provide heat hence …

(Image 185): … the sight of 73969 being piloted by Class 66/7 66737 *Leisa* as they awaited departure with the Inverness portion of the Highland Sleeper service from London Euston.

(Image 186): A visit was made to Edinburgh Waverley in the early hours of 22 July 2016 when 73967 (rebuilt from Class 73/0 73006) was piloted by Class 66/7 66705 *Golden Jubilee* as they arrived with the southbound Aberdeen portion of the Highland Sleeper service to London Euston.

(Image 187): Once released from their sleeper portion 73967 and Class 66/7 66705 *Golden Jubilee* prepare to move to Craigentinny T&RSMD for re-fuelling then …

(Image 188): … return to await the arrival of the Highland Sleeper service from London Euston to work the Aberdeen portion forward.

(Image 189): 73968 and 73966 (rebuilt from Class 73/0 73005) stand in Edinburgh Waverley in the early hours of 22 June 2016 as they await the arrival of the Highland Sleeper service from London Euston then …

(Image 190): … await the signal to depart with the Fort William portion. Note the pointwork allowing the train to depart whilst the Aberdeen portion further up the platform is prepared for departure.

(Image 191): The driver of 73971 checks his controls in the early hours of 19 July 2019 as he awaits the arrival of the Highland Sleeper service from London Euston.

A visit was made to Edinburgh Waverley in the early hours of 19 July 2019 when …

(Image 192): … 73971 arrived with the Aberdeen portion of the Highland Sleeper service to London Euston

(Image 193): … 73969 returned from re-fuelling at Craigentinny T&RSMD to reverse onto the northbound Fort William portion of the Highland Sleeper from London Euston.

(Image 194): … 73969 awaits the signal to depart with the northbound Fort William portion of the Highland Sleeper service from London Euston.

(Image 195): … 73971 awaits the signal to depart with the northbound Aberdeen portion of the Highland Sleeper service from London Euston.

Whilst dedicated to sleeper services north of Edinburgh, class members have seen occasional forays south of the border as on 18 March 2016 when 73969 and 73968 arrived at Carnforth to collect Royal Scotsman stock from the West Coast Railway Company's base following winter maintenance and storage.

Above (Image 196): 73969 and 73968 propel the Royal Scotsman trainset into the network sidings after collecting it from the West Coast Railway Company's base – originally Carnforth steam depot – then …

Right (Image 197): …. draw forward onto the main line en route to Kilmarnock where further stock was collected.

(Image 198): 73970 drifts past Balshaw Lane Junction on 5 June 2018 whilst en route from Crewe to Craigentinny T&RSMD after receiving attention at Wabtec/Brush Traction works at Loughborough.

SECTION 5: CLASS HB = E6101-6110; CLASS 74 = 74001-74010

When the Bournemouth Electrification Scheme was approved, it was also accepted that a more powerful electro-diesel locomotive was needed to power the numerous boat trains between London Waterloo and Southampton Docks and Weymouth with its Channel Islands Boat Train service. Rather than upgrade the Class 73 design it was decided to convert ten surplus Class HA, later Class 71, electric locomotives, rated at 2552hp, by the addition of a Paxman 6YJXL engine rated at 650hp at 1500 rpm. The selected locomotives, E501 5/16/06/00/19/23/03/05/17/21, became initially Class HB E6101-6110 respectively, later Class 74 74001-74010.

Despite hopes, the locomotives proved unreliable in service, hence their poor reliability, combined with a downturn in both boat train and newspaper/mail services, saw the locomotives, introduced to service between 1967-1968, became surplus to requirements, lasting only ten years before being withdrawn from service in December 1977.

(Image 199): The class was nominally allocated to Eastleigh to service the Bournemouth Line's newspaper/mail and boat train services to both Southampton Docks and Weymouth. This resulted in many being present on the depot as on 5 December 1976, when 74001 and 74010 were stabled nose to nose awaiting their next duty.

(Image 200): 74001 calls at Eastleigh on 29 December 1975 whilst working the Southampton Bevois Park-Clapham Junction empty van service.

(Image 201): 74003 awaits departure from London Waterloo on the evening of 28 October 1977 with a relief Channel Islands Boat Train that it will work as far as Bournemouth where a Class 33/1 will work the train forward. This was the last service for the 1977 season, hence the last to be worked by Class 74 traction (see Image 215).

(Image 202): 74001 awaits departure from London Waterloo on 8 October 1977 whilst working the 02:45 Waterloo-Poole newspaper/mail service which also comprised passenger accommodation.

(Image 203): 74002 stables in Eastleigh depot yard on 28 December 1975 awaiting its next duty.

(Image 204): 74004 undergoes maintenance in Eastleigh depot on 5 December 1976.

(Image 205): 74005 stands in Waterloo in the early hours of 8 September 1977 after arriving with empty vans from Clapham Junction to form an early morning newspaper/mail service.

(Image 206): 74005 awaits departure from London Waterloo in the early hours of 5 February 1977 whilst working the 02:15 Waterloo-Bournemouth newspaper/mail service which also included passenger accommodation.

(Image 207): 74007 stables in Eastleigh depot yard on 5 December 1976 whilst awaiting its next duty.

The first withdrawal came when 74006 caught fire during 1975 and was deemed beyond economic repair, being withdrawn from service in June 1976. After storage at Eastleigh (both depot and works) it was moved to George Cohen's Cransley scrapyard near to Kettering where it was cut up during June 1977.

(Image 208): 74006 lies stored on Eastleigh depot on 28 December 1975 awaiting a decision as to its future.

(Image 209): 74006 lies in George Cohen's scrapyard on 18 June 1977 as cutting up begins.

(Image 210): 74007 arrives at Eastleigh on 6 December 1976 whilst working the 08:23 Totton-Eastleigh commuter service; this service used the stock of the 02:15 London Waterloo-Bournemouth newspaper service which was subsequently marshalled into a van train for return to Clapham Junction ...

(Image 211): ... as on the previous day when 74007 awaited departure with the marshalled trainset en route to Clapham Junction.

(Image 212): 74008 awaits departure from London Waterloo on 5 February 1977 whilst working the 02:45 Waterloo-Poole newspaper/mail service which included passenger accommodation.

Right (Image 213): 74010 stables in London Waterloo in the early hours of 29 December 1975 whilst awaiting its next duty.

Below (Image 214): 74010 stables in Eastleigh depot yard on 5 December 1976 whilst awaiting its next duty.

(Image 215): 74007 accelerates out of Southampton Central on 28 October 1977 whilst working the last Channel Island Boat Train of the season. This proved not to be the case as noted at Image 201.

(Image 216): 74009 stands at the buffer stops in London Waterloo in the early hours of 5 February 1977 after bringing stock from Clapham Junction to form an early morning newspaper/mail service.

The finale of the class came in December 1977 and was marked with a 'Farewell' tour organised jointly by the Diesel & Electric Group and John Vaughan's Railway Pictorial Publications Railtours organisation that ran on 3 December 1977.

(Image 217): 74003 is seen here curving past the disused Southampton Terminal during its tour of Southampton Docks.

(Image 218): 74003 is having problems at Romsey where the diesel engine overheated and it proved difficult to re-start the engine. Whilst the delay allowed extra time for photography, it also incurred a one hour delay, a truncation of the scheduled itinerary and the replacement of 74003 by 74010 for the final Bournemouth-London Waterloo leg of the railtour.

(Image 219): 74003 stops for a photography session at Brockenhurst prior to continuing down the Lymington branch.

(Image 220): 74003 curves onto the Lymington branch after collecting the branch token whilst working the Brockenhurst-Lymington Pier leg of the 'Farewell' railtour.